Agile Illustrated: A Visual Learner's Guide to Agility
By Mike Griffiths, PMP, PMI-ACP

The Agile Manifesto

 Individuals and interactions over processes and tools

 Working software over comprehensive documentation

 Customer collaboration over contract negotiation

 Responding to change over following a plan

Contents

Chapter 1 – Introduction 4
   Scope 4
Chapter 2 - The Agile Manifesto Values and Principles 6
Chapter 3 – The Agile Manifesto Principles 10
Chapter 4 – The Declaration of Interdependence 24
Chapter 5 - PMI-ACP Domains 31
Domain 1 - Agile Principles and Mindset 33
Domain 2 – Value-Driven Delivery 40
Domain 3 – Stakeholder Engagement 49
Domain 4 – Team Performance 58
Domain 5 – Adaptive Planning 66
Domain 6 - Problem Detection and Resolution 74
Domain 7 – Continuous Improvement 79
Conclusion 85
Other Products 86
About the Author 88

Dedication

In memory of my mother, Sheila Griffiths, who was always curious and learning something new. Her passion for ideas was infectious. I miss our chats about everything from health to computers and am eternally grateful.

The Legal Stuff

Copyright © 2019 all rights reserved. Except as permitted under the United States Copyright Act of 1976, no part of this publication may be reproduced or distributed in any form or by any means, or stored in a database or retrieval system, without the prior written permission of the publisher.

"PMI-ACP," "PMI Agile Certified Practitioner (PMI-ACP)," and "PMI Registered Education Provider (R.E.P)," are registered marks of the Project Management Institute, Inc.

Chapter 1 – Introduction

Welcome to ***Agile Illustrated: A Visual Learner's Guide to Agility***. This book shows you how agile works by using graphics to illustrate the core components of the agile mindset and agile team behaviors.

This book is for fellow visual thinkers who like to see the big picture before getting into details. Sometimes called "right-brained," after the portion of the brain responsible for processing images, we would rather see how something works than being told the information in detail.

Research into visual thinking by David Hyerle, creator of Thinking Maps methodology, reports that 90% of the information entering the brain is visual. Forty percent of all nerve fibers connected to the brain are connected to the retina and a full 20% of the entire cerebral cortex is dedicated to vision, so let's use it.

By using mind maps, this book shows the workings of agile teams and illustrates how all the parts link together. Many memorization techniques are based on spatial memory: By associating ideas with places, we can recall them more easily. Using a combination of images, mind maps, and explanations, we engage the right and left hemispheres of our brains to build stronger comprehension and better recall.

Scope
This book covers the 4 values of the Agile Manifesto, the 12 Agile Principles, and the 6 principles of the Project Management Declaration of Interdependence. For anyone studying for the PMI-ACP® exam, it also covers the 7 Domains and all 62 of their associated tasks. This coverage is depicted on the next page:

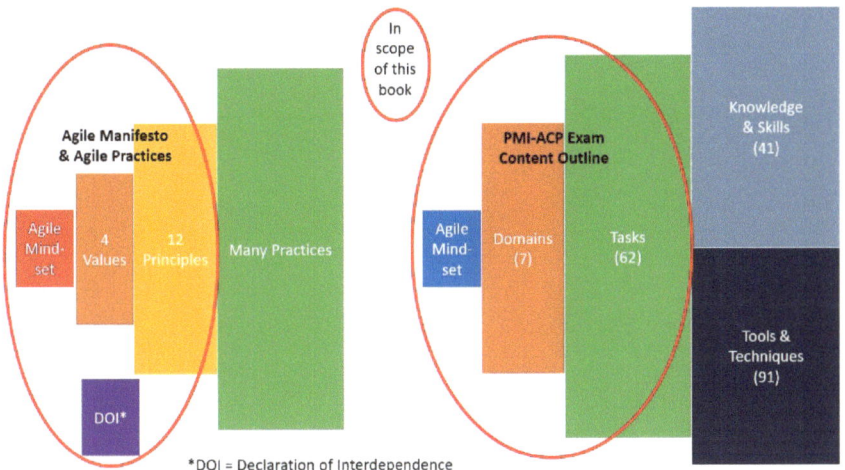

Notice that this book does not cover the associated Knowledge and Skills or Tools and Techniques that fall under the tasks. So, it will not provide everything you need to know to pass the PMI-ACP exam – for that I recommend my *PMI-ACP® Exam Prep* book.

However, it will explain the higher-level values, principles, and tasks that are tested in the exam. Plus, it does so in a highly visual format with lots of mind maps and graphics to *show* you what you need to know, rather than a regular textbook that *tells* you what you need to know (if you read it all and can then recall it).

Chapter 2 - The Agile Manifesto Values and Principles

The Agile Manifesto

 **Individuals and interactions over processes and tools**

 **Working software over comprehensive documentation**

 **Customer collaboration over contract negotiation**

 **Responding to change over following a plan**

The Agile Manifesto was created during a meeting in February 2001 that brought together a number of software and methodology experts who were in the forefront of the emerging agile methods. Let's look at the values one by one.

 **Individuals and interactions over processes and tools**

Value 1 – Individuals and Interactions over processes and tools

While processes and tools will likely be necessary, we should try to focus attention on the individuals and interactions involved. This is because work is undertaken by people, not tools, and problems get solved by people, not processes. Likewise, products are accepted by people, scope is debated by people, and the definition of a successfully "done" project is negotiated by people.

What will help set up a project for success is an early focus on developing the individuals involved and an emphasis on productive and effective interactions. Processes and tools can help, yet projects are ultimately about people. So, to be successful, we need to spend the majority of our time in what may be the less comfortable, messy, and unpredictable world of people.

Working software over comprehensive documentation

Value 2 – Working software over comprehensive documentation

This value speaks to the need to deliver. It reminds us to focus on the purpose or business value we're trying to deliver, rather than on paperwork.

Many developers are detail-oriented and process-driven. While these characteristics are often highly beneficial, they can also mean the developer's focus is easily distracted from the real reason they are undertaking software projects—to write valuable software. So, this emphasis on valuing working software over comprehensive documentation acts as a useful reminder of why these projects are commissioned in the first place—to build something useful. Documentation by itself, or at the expense of working software, is not useful.

 **Customer collaboration over contract negotiation**

Value 3 – Customer collaboration over contract negotiation

We need to be flexible and accommodating rather than fixed and uncooperative. This involves tradeoffs between the development team and business rather than reverting back to contracts and statements of work. We could build the product exactly as originally specified, but if the customer's preferences or priorities change, it would be better to be flexible and work toward the new goal.

It is difficult to define an up-front, unchanging view of what should be built. This challenge stems from the dynamic nature of knowledge work products, especially software systems. Software is intangible and difficult to reference: companies rarely build the same systems twice, business needs change quickly, and technology changes rapidly.

We should recognize at the start that things are going to change, and we'll need to work with the customer throughout the project to reach a shared definition of "done." This requires a more trusting relationship and more flexible contract models than we often see on projects.

 **Responding to change over following a plan**

Value 4 – Responding to change over following a plan

The quote from scholar Alfred Korzybski, "The map is not the territory," warns us not to follow maps if they do not match the surroundings. Instead, trust what you see and act accordingly.

In modern, complex projects, we know our initial plans will likely be inadequate. They are based on insufficient information about what it will take to complete the project.

Agile projects have highly visible queues of work and plans in the form of backlogs and task boards. The intent of this value is to broaden the number of people who can be readily engaged in the planning process by adjusting the plans and discussing the impact of changes.

Chapter 3 – The Agile Manifesto Principles

The authors of the Manifesto also created twelve guiding principles for agile approaches:

Agile Manifesto Principles

 Our highest priority is to satisfy the customer through early and continuous delivery of valuable software.

 Welcome changing requirements, even late in development. Agile processes harness change for the customer's competitive advantage.

 Deliver working software frequently, from a couple of weeks to a couple of months, with a preference to the shorter timescale.

 Business people and developers must work together daily throughout the project.

 Build projects around motivated individuals. Give them the environment and support they need, and trust them to get the job done.

 The most efficient and effective method of conveying information to and within a development team is face-to-face conversation

 Working software is the primary measure of progress.

 Agile processes promote sustainable development. The sponsors, developers, and users should be able to maintain a constant pace indefinitely.

 Continuous attention to technical excellence and good design enhances agility.

 Simplicity–the art of maximizing the amount of work not done–is essential.

 The best architectures, requirements, and designs emerge from self-organizing teams.

 At regular intervals, the team reflects on how to become more effective, then tunes and adjusts its behavior accordingly.

Now let's examine the 12 agile principles:

1 – Our highest priority is to satisfy the customer through early and continuous delivery of valuable software.

This principle reminds us to satisfy the customer via early and continuous delivery of value. We must structure the project and the development team to deliver value early and frequently.

We must also remember that what we are delivering is valuable software, not completed work products such as a WBS, documentation, or plans. We need to stay focused on the end goal. For software projects, this is the software; for other types of projects, the end goal will be the product or service we have been asked to deliver or enhance.

2 – Welcome changing requirements, even late in development. Agile processes harness change for the customer's competitive advantage.

Changes can be good for a project if they allow us to deliver a late-breaking, high-priority feature. Yet in non-agile projects, changes are often seen as negative; they may be considered "scope creep" or blamed for the project deviating from the plan.

Agile approaches favor lightweight, high-visibility approaches for managing change; for example, continuously updating and prioritizing changes into the backlog of work to be done. Agile's well-understood, high-visibility approach for handling changes keep the project adaptive and flexible as long as possible.

3 – Deliver working software frequently, from a couple of weeks to a couple of months, with a preference to the shorter timescale.

This principle emphasizes the importance of releasing work to a test environment and getting feedback quickly. Agile teams need feedback on what they have created thus far to see if they can proceed, or if a change of course is needed.

Delivering within a short timeframe also has the benefit of keeping the product owner engaged and keeping dialogue about the project going. With frequent deliveries, we will regularly have results to show the customer and opportunities to get feedback. Often at these demos, we learn of new requirements or changes in business priorities that are valuable planning inputs.

It's human nature to want our work to be as perfect as possible before sharing it. However, we are doing ourselves a disservice by holding on to our work for so long. It's better to get feedback early and often to avoid going too far down the wrong track.

4 – Business people and developers must work together daily throughout the project.

The frequent demos mentioned in principle 3 are one example of how business people and developers work together throughout the project. Daily face-to-face engagement with the customer is one of the most difficult principles to ensure from a practical standpoint, but it is worth pushing for. Written documents, e-mails, and even telephone calls are less efficient ways of transferring information than face-to-face interactions.

By working with business representatives daily, we can learn about the business in a way that is far beyond what a collection of requirements-gathering meetings can ever achieve. As a result, we are better able to suggest solutions and alternatives to business requests. The business representatives also learn what types of solutions are expensive or slow to develop, and what features are cheap. They can then begin to fine-tune their requests in response.

When it isn't possible to have daily interactions between the business representatives and the development team, agile approaches try to get the two groups working together regularly in some way, perhaps every two days or whatever type of frequent involvement will work. (Some teams use a "proxy customer," in which an experienced business analyst [BA] who is familiar with the business interests serves as a substitute, but this isn't an ideal option.)

5 – Build projects around motivated individuals. Give them the environment and support they need, and trust them to get the job done.

It is significantly more important for a project to have the best people possible than to have the best processes and tools. So, making sure smart and motivated people are on the team is likely to make a big difference in whether our project is delivered successfully and efficiently.

While we may not always be able to pick our dream team, we can motivate and empower the team members we do have. Agile approaches promote empowered teams. People work better when they are given the autonomy to organize and plan their own work. Agile approaches advocate freeing the team from the micromanagement of completing tasks on a Gantt chart. Instead, the emphasis is on craftsmanship, peer collaboration, and teamwork, which result in higher rates of pride and productivity.

Knowledge work projects involve team members who have unique areas of expertise. Such people do their best work when they are allowed to make their own decisions and local planning for the project.

For leaders, this doesn't mean abdicating involvement or abandoning the team to fend for itself; instead, we recognize that our team members are experts in what they do, and we provide the support they need to ensure they are successful.

6 – The most efficient and effective method of conveying information to and within a development team is face-to-face conversation.

Written documents are great for creating a lasting record of events and decisions, but they are slow and costly to produce. In contrast, face-to-face communication allows us to quickly transfer a lot of information in a richer way that includes emotions and body language.

Of course, face-to-face conversations can't be applied to all project communications, but agile teams aim to follow it whenever possible. This is one example of how agile methods need to be customized or scaled for each project. As team sizes grow, it becomes harder to rely on face-to-face communications and an appropriate level of different-format communications need to be introduced.

7 – Working software is the primary measure of progress.

By adopting "working software" (or "working systems") as a primary measure of progress, our focus is shifted to working results rather than documentation and design. In agile, progress is assessed based on the emerging product or service we are creating. Questions like "How much of the solution is done and accepted?" are preferred over "Is the design complete?" since we want to focus on usability and utility rather than conceptual progress.

This definition of progress as "working systems" creates a results-oriented view of the project. Interim deliverables and partially completed work will get no external recognition. We want to instead focus on the primary goal of the project—a product that delivers value to the business.

8 – Agile processes promote sustainable development. The sponsors, developers, and users should be able to maintain a constant pace indefinitely.

Agile methods strive to maximize value over the long term. Some of the rapid application development (RAD) techniques that preceded agile promoted—or at least accepted—intense periods of prototyping prior to demos. Yet, teams working long hours over an extended period of time results in burn-out and mistakes. This is not a sustainable practice.

Instead of long, intense development periods, agile approaches favor a sustainable pace that allows team members to maintain a work-life balance. A sustainable pace is not only better for the team; it benefits the organization as well. Long workdays lead to resignations, which means the organization loses talent and domain knowledge. Hiring and integrating new members into a team is a slow and expensive process.

Instead, working at a pace that can be maintained indefinitely leads to a happier and more productive team. Happy teams also get along better with business representatives than do overworked teams. There is less tension and work relationships improve.

9 – Continuous attention to technical excellence and good design enhances agility.

An agile team needs to balance its efforts to deliver high-value features with continuous attention to the design of the solutions. This balance allows the product to deliver long-term value without becoming difficult to maintain, change, or extend. Preventative maintenance and cleaning up code are preferable to fixing problems. This helps the project run more smoothly and speeds up the team's progress.

In the software world, once the code base becomes too messy or tangled, the organization loses its ability to respond to changing needs. In other words, it loses its agility. So we need to give the development team enough time to undertake refactoring. Refactoring is the housekeeping, cleanup, and simplification that needs to be made to code in order to ensure it is stable and can be maintained over the long term.

10 – Simplicity ☐ the art of maximizing the amount of work not done ☐ is essential.

The most reliable features are those we don't build—since there is nothing that could go wrong with them. In the software world, up to 60 percent of features that are built are used either infrequently or never. Because so many features that are built are never actually used, and because complex systems have an increased potential to be unreliable, agile approaches focus on simplicity. This means boiling down the requirements to their essential elements only.

Complex projects take longer to complete, are exposed to a longer horizon of risk, and have more potential failure points and opportunities for cost overruns. Therefore, agile methods seek the "simplest thing that could possibly work" and recommend that this solution be built first. This approach is not intended to preclude further extension and elaboration of the product, instead, it simply says, "Let's get the plain-vanilla version built first." This approach not only mitigates risk but also helps boost sponsor confidence.

11 – The best architectures, requirements, and designs emerge from self-organizing teams.

To get the best out of people, we need to let them self-organize. It allows people to find an approach that works best for their methods, their relationships, and their environment. They will thoroughly understand and support the approach because they helped create it. As a result, they will produce better work.

Self-organizing teams that have the autonomy to make local decisions have a higher level of ownership and pride in the architectures, requirements, and designs they create than in those forced upon them or "suggested" by external sources. Ideas created by the team have already gone through the team vetting process for alignment and approval, so they don't need to be "sold" to the team. In contrast, ideas that come from outside sources need to be sold to the team for the implementation to be successful, and this is sometimes a challenging task.

Another factor that supports this principle is that the members of a self-organizing project team are closest to the technical details of the project. As a result, they are best able to spot implementation issues, along with opportunities for improvements. So instead of trying to educate external people about the evolving structure of the project, agile approaches leverage the capacity of the team to best diagnose and improve the project's architectures, requirements, and designs. After all, the team members are the most informed about the project and have the most vested in it.

**12 – At regular intervals, the team reflects on how to become more effective, then tunes and adjusts
its behavior accordingly.**

Gathering lessons learned at the end of a project is too-little, too-late. Instead, we need to gather lessons learned while they are still applicable and actionable. This means we need to gather them during the project and, most importantly, make sure we do something about what we've learned in order to adjust how we complete the remainder of the work.

Agile approaches employ retrospectives to reflect on how things are working and identify opportunities for improvements. These retrospectives are done at the end of each iteration, ensuring that the team has regular opportunities to review their process. An advantage of doing retrospectives so frequently is that we don't forget about problems and issues. Compare this to conducting a single lessons learned review at the end of a project, in which team members are asked to think back over a year or more to recall what went well and where they ran into issues.

Another disadvantage of only gathering lessons learned at the end of a project is that the lessons won't really be helpful to the organization until another project with a similar business or technical domain or team dynamics comes along. At that point, it is easy to dismiss the lessons learned from an earlier project as not applicable to the current situation. On an agile project, we capture the lessons learned as we progress, so we can't pretend they aren't applicable. We know they are relevant, and we are motivated to tune and adjust our process accordingly.

Chapter 4 – The Declaration of Interdependence

A lesser-known cousin to the Agile Manifesto, the Declaration of Interdependence was created in 2005 to describe management principles required to achieve an Agile Mindset in product and project management. It describes six principles essential to modern project management. In this chapter, we will review them one by one.

Declaration Of Interdependence

We Increase return on investment by making a continuous flow of value our focus.

We Deliver reliable results by engaging customers in frequent interactions and shared ownership.

We Expect uncertainty and manage for it through iterations, anticipation and adaptation.

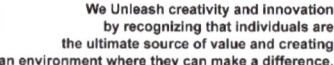

We Unleash creativity and innovation by recognizing that individuals are the ultimate source of value and creating an environment where they can make a difference.

We Boost performance through group accountability for results and shared responsibility for team effectiveness.

We Improve effectiveness and reliability through situationally specific strategies, processes and practices.

1 – We increase return on investment by making continuous flow of value our focus.

Amaze your customers; keep giving them what they ask for!

Concentrate on developing features the business asks for: This is how we can get the best benefits for the business and support for the process. Projects are hard to cancel or deny requests from when they consistently deliver business results.

2 – We deliver reliable results by engaging customers in frequent interactions and shared ownership.

When planning interaction with the business, try to be more like the good neighbor you see frequently and can easily call upon rather than the intrusive relative who moves in for a while and then disappears for a year. We want regular and engaging business interaction, not a huge, upfront requirements-gathering phase followed by nothing until delivery. Frequently show how the system is evolving and make it clear the business drives the design by listening to and acting on feedback.

3 – We expect uncertainty and manage for it through iterations, anticipation, and adaptation.

Since software functionality is hard to describe, technology changes quickly and so do business needs. Software projects typically have lots of unanticipated changes. Rather than trying to create and follow a rigid plan that is likely to break, it is better to plan and develop in short chunks (iterations / sprints) and adapt to changing requirements.

4 – We unleash creativity and innovation by recognizing that individuals are the ultimate source of value, and creating an environment where they can make a difference.

We manage property and lead people; if you try to manage people they feel like property.

Projects are completed by living, breathing people, not tools or processes. To get the best out of our team we must treat them as individuals, provide for their needs and support them in the job. Paying a wage might guarantee that people show up, but how they contribute once they are there is governed by a wide variety of factors. If you want the best results, provide the best environment you can.

5 – We boost performance through group accountability for results and shared responsibility for team effectiveness.
Everyone needs to share responsibility for making the project, and the team as a whole, successful. We can help by empowering the team to make their own decisions. When people are more engaged in a process, they are more committed to its outcome and success. In short, people care more about things they had a hand in creating than things given to them or imposed upon them.

6 – We improve effectiveness and reliability through situationally specific strategies, processes and practices.

Real projects are complex and messy. Rarely do all the ideal conditions for agile development present themselves. Instead, we have to interpret the situation and make the best use of the techniques, people, and tools available to us. There is no single cookbook for how to run successful projects; instead we need to adjust to best fit the project ingredients and project environment we are presented with.

Chapter 5 - PMI-ACP Domains

The Project Management Institute (PMI)® Agile Certified Practitioner (PMI-ACP)® exam tests knowledge and skills across seven domains. Descriptions of these domains, the tasks they contain, and the percentage of questions from each domain are summarized below.

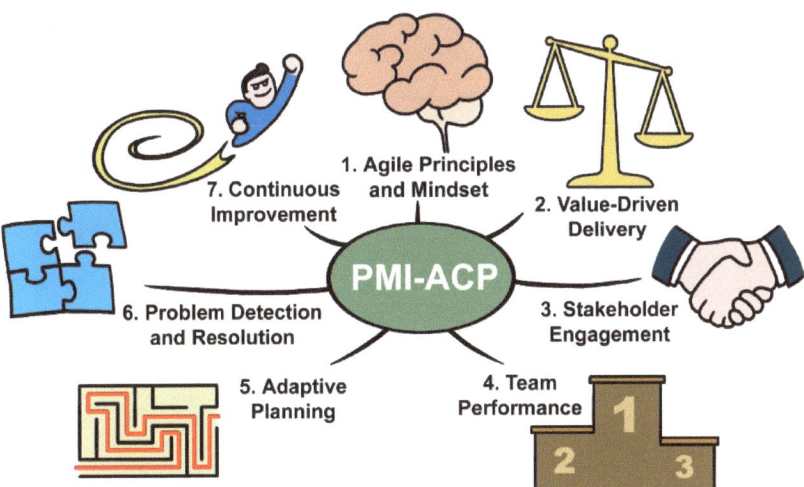

The number of exam questions in each category within these domains are broken out as follows:

PMI-ACP Exam Item Count

Domain	% of Q.	# of Q.
1. Agile Mindset	16%	19
2. Value-Driven	20%	24
3. Stakeholder Engagement	17%	20
4. Team Performance	16%	19
5. Adaptive Planning	12%	14
6. Problem Resolution	10%	12
7. Continuous Improvement	9%	11
		120

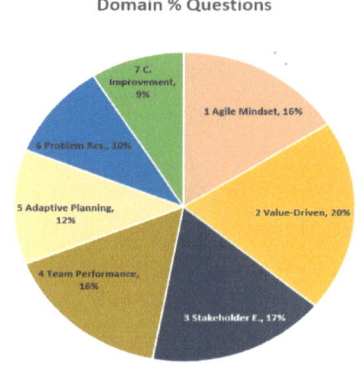

We will now examine each domain and learn about the tasks within that domain.

Domain 1 - Agile Principles and Mindset

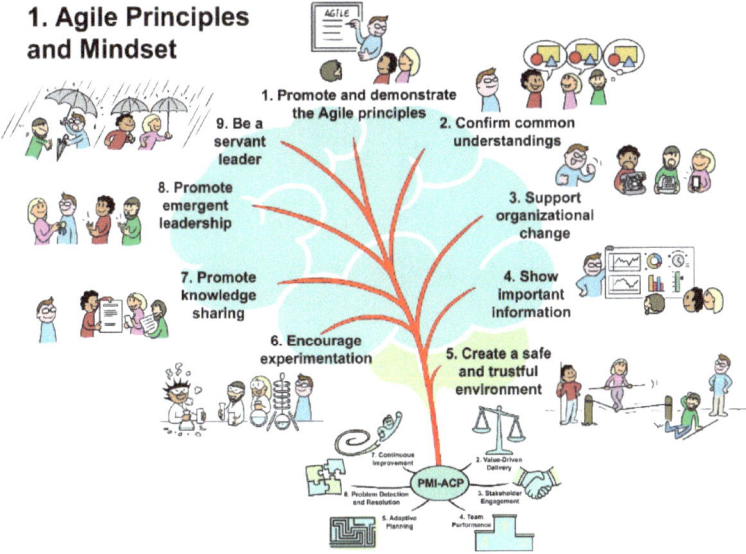

The Agile Principles and Mindset domain is all about exploring, embracing, and applying the agile principles and mindset within the context of the project team and organization. It contains 9 tasks. Understanding and applying the tasks accounts for 16%, or around 18-20 of the PMI-ACP exam questions.

Task 1 – Promote and demonstrate the Agile principles

Share and explain agile ideas to others. Behave as you would like others to act.

By explaining and demonstrating the values and principles that make up the agile mindset, we help others understand and adopt them. By talking about the agile values within our teams and with our customers, we create a better shared understanding of what it means to be agile.

Task 2 – Confirm common understandings

Make sure everyone shares the same view and mental model.

We should confirm everyone has an aligned view of agile principles, how we work, and the words we use to describe things.

Task 3 – Support organizational change

Help make things better in the organization.

We should help improve the systems and processes in our organization. This includes education and influencing to make things more efficient and effective.

Task 4 – Show important information

Keep important data and metrics visible to everyone.

It is not enough to have useful information, it needs to be displayed, visually and prominently. Visualizing real performance improves transparency, buy-in, and trust.

Task 5 – Create a safe and trustful environment

Create a safe team space where people feel comfortable experimenting, failing, and learning. This way people can improve without fear of ridicule or judgement.

Task 6 – Encourage experimentation

Encourage people to try new approaches and techniques then learn from them. Encouraging experimentation accelerates learning and innovation. We learn from failure, too, so try new things, capitalize on what works and learn from what does not work. Innovate and improve.

Task 7 – Promote knowledge sharing

Encourage people to share what they learn ☐ good and bad.

We need to share what we learn so others can benefit from our knowledge. Likewise, we need to pick up what others have learned so we stay current, productive, and prevent that knowledge resides in just a few key people.

Task 8 – Promote emergent leadership

Help people step up to take on more responsibility.

Create a respectful workplace where people can take on new work and roles progressively and safely. Encourage people to improve and recognize all efforts to do so, whether they are immediately successful or not.

Task 9 – Be a servant leader

Serve the team, clear obstacles from their path, and make sure they have what they need.

People need encouragement and support to try new things and to deliver in challenging environments. Provide what they need, whether that's help with a new tool, an introduction to a customer, or just some kind words of encouragement. Help make them successful as best you can.

Domain 2 – Value-Driven Delivery

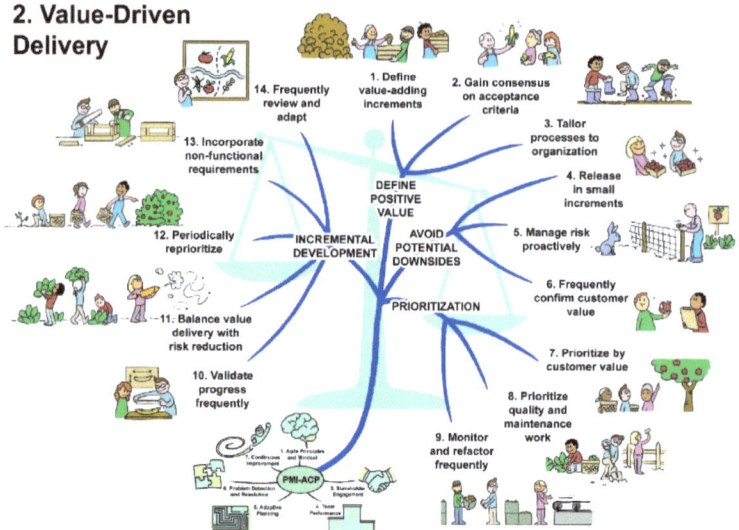

The Value-Driven Delivery domain covers defining positive value, avoiding potential downsides (loss of value), prioritization, and incremental development. Understanding and applying the tasks in this domain account for 20%, or around 24 of the PMI-ACP exam questions.

Define Positive Value

Task 1 – Define value-adding increments

Break work into small chunks that we can more easily undertake and measure progress against.

Look for tasks that add value and maximize those. Also look for things that do not add value and aim to minimize those things. This way we can deliver the most value to the business/customer.

Task 2 – Gain consensus on acceptance criteria

Make sure everyone knows what "Done" looks like.

Understand, discuss, and share what success and acceptable is and what it is not. Make sure people know in a timely manner what we are aiming for, so expectations are aligned.

Task 3 – Tailor processes to organization

Change how we work to match our environment.

Engage the team in determining and tailoring how everyone works to best fit the organizational characteristics they face. Make sure processes fit the situation and we are optimizing the value delivered.

Avoid Potential Downsides

Task 4 – Release in small increments

Ship product as soon as it is ready, in small batches.

Identify minimally marketable features and minimally viable products to determine the smallest useful units of functionality. Build, release, and get feedback on these small chunks to quickly gain feedback, learn, adapt, improve, and deliver value.

Task 5 – Manage risk proactively

Actively look for issues and problems and try to prevent them before they occur.

Getting feedback early on small increments of the product helps us avoid large and costly changes later.

Task 6 – Frequently confirm customer value

Frequently check with the customer to make sure they are okay and getting what they need.

Engage the customer as frequently as you can; they often spot things development teams overlook and they are the ultimate accepters/rejecters of the product.

Prioritization

Task 7 – Prioritize by customer value

Order the work by what the customer wants the most.

Collaborate with the business and customers to find out what their priorities are. Wherever practical, order work based on this sequence. We want to deliver the highest value items as soon as we can – aware of dependencies.

Task 8 – Prioritize quality and maintenance work

Make sure we do not forget quality and maintenance. Do this work along with the regular development and delivery of product increments.

Frequently review to see if enhancements or quality work is necessary. Doing things sooner rather than later is usually the most economic.

Task 9 – Monitor and refactor frequently

Frequently review how we are working and incorporate any improvements we have identified.

Mix quality improvement and preventative maintenance work in with regular development. Think "daily hygiene" not "spring cleaning," we need to be doing this work as we go, not save it up for improvement efforts. This improves our ability to respond to changes – enhancing agility.

Incremental Development

Task 10 – Validate progress frequently

Frequently check progress to make sure we are on track and quality is still good.

If things need to change, we want to learn about them sooner rather than later to reduce the amount of scrap or rework that may be required.

Task 11 – Balance value delivery with risk reduction

Do both production and protecting the production system.

We achieve most over the long-term when we monitor both the product and the system that produces it. Look for threats and proactively avoid or reduce them. This work is important, so put this work in the backlog, estimate it, and plan it just as we would other work.

Task 12 – Periodically reprioritize

Every now and again check that work is being done on the right things in the right quantities.

As change occurs, make sure we are still focusing on the most important things. Regularly pause, evaluate, and rebalance or adjust our focus as required to maximize delivered value.

Task 13 – Incorporate non-functional requirements

Do maintenance work as well as production work.

Make sure to think beyond just our product and look into the environment in which it will be used. Make sure we also prioritize work that will help make our product and its customers successful. Ultimately, it is this combination and product and environment that will create a success or a failure. So do not ignore these tasks.

Task 14 – Frequently review and adapt

Look at the whole, rebalance work streams, try new ideas, and improve the whole.

Review what we are building and collect improvement suggestions. See if we can make changes to improve the overall product/service and process.

Domain 3 – Stakeholder Engagement

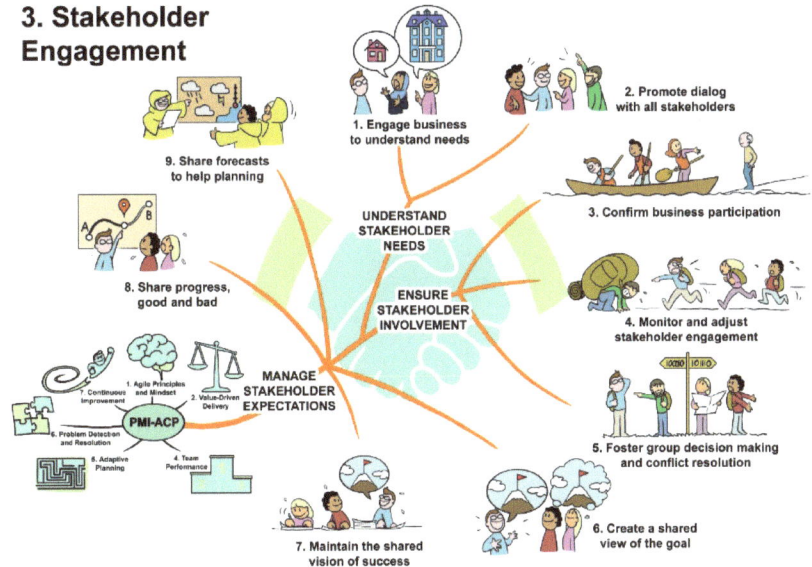

The Stakeholder Engagement domain deals with understanding stakeholder needs, ensuring stakeholder involvement, and managing stakeholder expectations. It accounts for 17% of the questions, which means you can expect to get around 20 questions on the exam on this topic.

Understand Stakeholder Needs

Task 1 – Engage business to understand needs

Talk to people to understand what they really want and expect from the project.

Often there is a difference between what people initially describe and what they really want. In addition, once people see "X" they often want "Y." So it is important to explore what they really mean when describing something and check
 back periodically to see if their needs are still being met.

Task 2 – Promote dialog with all stakeholders

Encourage all participants to share their thoughts, ideas, and concerns.

Make sure you know who all your customers are. Engage them early and often to keep the dialog going. We want to avoid nasty surprises and disappointments, so keep communicating.

Ensure Stakeholder Involvement

Task 3 – Confirm business participation

Get agreement on the type and amount of involvement that will be required from the business for this product/project to be successful.

Discuss what is expected for business participation. Agree on how frequently business involvement is required, what form it will take, and how the process for evaluating it will operate.

Task 4 – Monitor and adjust stakeholder engagement

Continually check how work with people on the project is going.

As necessary, change how we are working with the business to better suit the work being done and address issues that are identified. Look for people who are overloaded with work and creating bottlenecks to throughput of value delivery. Rebalance workloads if necessary to optimize the flow of value.

Task 5 – Foster group decision making and conflict resolution

Encourage people to make decisions collectively and collaboratively.

Make sure to engage the relevant people, and that the process for making decisions taps individual knowledge and the group consensus for the best way forward. Getting ideas for solutions individually first avoids the halo effect (being drawn to a senior person's opinion) and groupthink or social loafing (also called the Ringelmann effect, where people do not try as hard when in a group.) Then as a group, decide on the best options presented.

Manage Stakeholder Expectations

Task 6 – Create a shared view of the goal

Unite people with a common view of where we are trying to get.

It is important that people understand and share the same vision of "Done" for increments and the final solution. To be effective and work productively without detailed, centralized coordination, we need people moving towards common goals. So, spend time ensuring everyone knows where we are trying to get to. That way, when faced with their own local decision points, or forks in the trail towards project completion, they make decisions aligned with the larger goal.

Task 7 – Maintain the shared vision of success

Frequently remind people about where we are trying to get to and why it is important that we get there.

Product and project priorities can change quickly. People come and go from projects and the market is constantly evolving. With all these changes occurring it is important that we frequently remind people of where we are trying to get to and why that is important. So, don't let the end goal shift or become fuzzy in people's minds. Align their expectations towards the success criteria to keep the end goal clearly in focus.

Task 8 – Share progress, good and bad

Share information. Make sure people know what is going on, whether this is good news or bad news.

It is very important to be open and honest about progress, issues, and threats. People are astute and recognize if things are not being discussed; they will begin to withdraw their wholehearted commitment if they feel information is being withheld. So share information, both good and bad with the team and business representatives ☐ often people surprise us with novel solutions to problems. People want to know what is going on and we should be able to discuss topics openly.

Task 9 – Share forecasts to help planning

By sharing true updates on progress, we help people plan their work and improve their ability to plan for the future.

When sharing forecasts and plans, we also need to share our levels of uncertainty. Future predictions are more useful when we also know the uncertainty connected to them.

Domain 4 – Team Performance

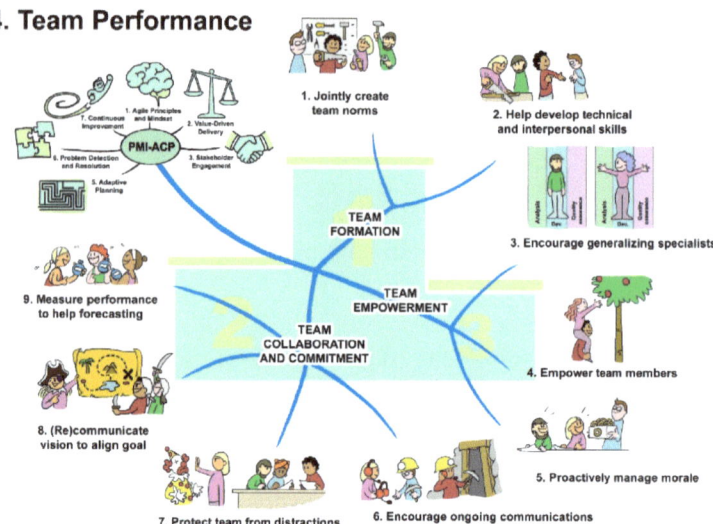

The Team Performance domain includes Team Formation, Team Empowerment, and Team Collaboration activities. Questions in this domain make up 16% of the total exam so you can expect 18-20 questions on the topic in the exam.

Team Formation

Task 1 – Jointly create team norms

Learn how people want to work and agree on how things should be done and how issues should be handled.

As a group, develop the group rules that will be followed. By being involved in the creation of the team norms, people are much more likely to feel ownership and commitments towards them. Telling people how we should work is much less empowering than engaging those people in jointly developing their own framework.

Task 2 – Help develop technical and interpersonal skills

Encourage the development of technical and people skills so everyone is equipped to work effectively.

Knowledge work requires two sets of skills. The first is to do the technical work as a subject matter expert (SME), the second is to work productively with other SMEs and stakeholders, including the business and customer. The job of learning and honing these skills is never done, and we should always be improving our technical and collaboration skills.

Team Empowerment

Task 3 – Encourage generalizing specialists

Encourage people to have a broad range of skills, not only deep, narrow ones, so that as workload varies people can help other team members out.

The concept of "T" shaped people rather than "I" shaped people captures the idea of having skills in surrounding fields of work, in addition to a specialization. To maximize the value delivered we want global rather than local optimization. This means focusing on overall throughput of value over people-utilization efficiencies. T-shaped people are valuable for optimizing value since they allow us to share work to reduce bottlenecks.

Task 4 – Empower team members

Encourage people to step up for new roles. Allow them to make their own decisions. Put them in charge of many elements of their job.

We want people to take ownership of their work and start to make their own improvements. So encourage people to look for opportunities for improvements and take initiative to make them happen. These are forms of emergent and shared leadership. Subject matter experts know their domains best, so empower them to manage complexity and create solutions to the problems they face.

Task 5 – Proactively manage morale

Learn what motivates people and provide that motivation in their workplace.

Frequently observe and ask team members about what motivates them individually and as part of a team. Also learn what demotivates or upsets them. Then try to find ways to improve the work environment to foster happiness, productivity, and satisfaction.

Team Collaboration and Commitment

Task 6 – Encourage ongoing communications

Encourage dialog and technology that helps share information.

Usually the best way to help communications is to physically co-locate with the people you need to communicate with. Nothing beats seeing them and talking with them. It allows for the richest exchange of information accompanied with body language and emotion.

When colocation is not possible, provide the best tools you can to keep people in communication. Monitor communications and look for ways to reduce miscommunication or address missing communication. This helps reduce costly and wasteful rework caused by miscommunication.

Task 7 – Protect team from distractions

Shield the team from interruptions.

Distractions and low-priority interruptions can come from many sources. They might be requests from superfluous sources or demands for low-priority admin work. Even quick interruptions cause task-switching and interrupting flow.

Special-ops and Skunkworks teams have been effective and highly productive in part because they were separated and shielded from interruptions.

Task 8 – (Re)communicate vision to align goal

Show the end goal and how people's contributions help get us there.

People should understand how their work contributes towards the end goal. So we need to align the team goals with the product or project goal and show the connections and steps along the way to our final destination.

Task 9 – Measure performance to help forecasting

Encourage people to measure and share their performance so we can get better at forecasting at a high level.

In order to improve our ability to forecast, we need to track how things actually turned out. If we keep making estimates without checking actual performance, we will keep making the same estimation errors. Tracking velocity and work delivered helps create a more accurate view of the team's true capacity for future work.

Domain 5 – Adaptive Planning

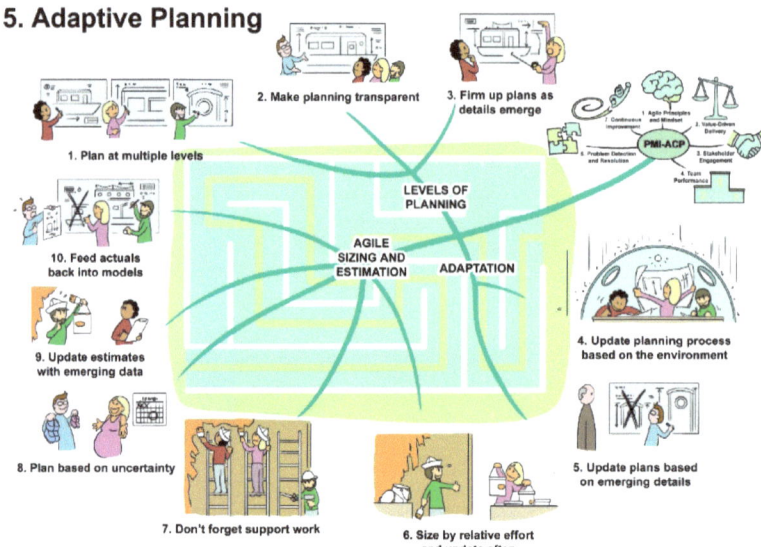

The Adaptive Planning domain covers different levels of planning, adaptation of plans, and agile sizing and estimation. This domain represents 12% of the exam questions, so expect to see around 14 questions on the domain in the exam.

Levels of Planning

Task 1 – Plan at multiple levels

Plan programs, projects, releases, features, and stories.

We need to plan various levels of detail. From big chunks like projects and releases down to small chunks such as features and stories. Initially, we may not have much information to go on, but then as the project progresses and we learn more, we can refine our plans based on emerging information.

Task 2 – Make planning transparent

Share your plans and work.

There should be no secrets and it is fine to share the limitations of our plans. Maybe someone will contribute some new information that helps us or points out a potential issue before we encounter it. So make plans visible and engage involvement by key stakeholders. This also helps gain buy-in and commitment towards meeting the plans.

Task 3 – Firm up plans as details emerge

Fill in the blanks as you get more information.

As the project progresses, we should update our plans and the planning process. This helps manage stakeholder expectations and build a shared understanding of deliverables and schedule.

Adaptation

Task 4 – Update planning process based on the environment

Adjust approach to planning based on the environment.

We should ensure our planning approach is appropriate for our project environment. So assess factors like the criticality, complexity, and size of the project, and adopt a planning approach that fits our context.

Task 5 – Update plans based on emerging details

Adjust the plan based on what happens.

Stuff changes, things happen, and we need to adapt. As new requirements emerge and parameters such as budget, schedule, and priorities change, make sure to update the plans. We aim to maximize the value being delivered and this demands staying current and on top of changes.

Agile Sizing and Estimation

Task 6 – Size by relative effort and update often

Make comparisons to known things; don't try to guess at absolute values.

People are generally better at estimating relative to known quantities than trying to determine absolute values. For example, it is easier for us to determine that Bill is taller than Ivan than to estimate Bill is 1.8 meters tall. Likewise, estimating relative to work already done tends to be more accurate and quicker than estimating the effort for each new piece of work.

Refine and improve estimates as you go, factoring what you learn about how things actually pan out.

Task 7 – Don't forget support work

In addition to creating features, there will be support and maintenance work that also needs estimating.

We need to ensure we do not estimate only initial development work but also consider any support, operations, or maintenance work that needs to be accounted for. We also need to adjust the available capacity of the team and schedules, budget, etc., to make allowances for this work.

Task 8 – Plan based on uncertainty

Don't be overly precise if the circumstances are vague.

We likely need an initial estimate of cost ranges, schedule, and scope to get an initiative approved. This initial plan is likely to be our least accurate since it is created when we know least about the project ☐ at the beginning. Be transparent about the levels of uncertainty inherent at that stage.

Task 9 – Update estimates with emerging data

As we receive new data, update the plans.

Update the estimates and plans for cost, scope, and schedule as details emerge. These estimates are today's best estimate that reflect our current understanding. Regularly updating our estimates helps us manage the project.

Task 10 – Feed actuals back into models

Add the results into the process.

It is important to feed the actuals back into the plans and reevaluate likely outcomes. These include details about actual levels of staff availability, the time taken to build things, defect rates, delays with approvals, etc. We need to incorporate all this information to build and maintain estimates to complete.

Domain 6 - Problem Detection and Resolution

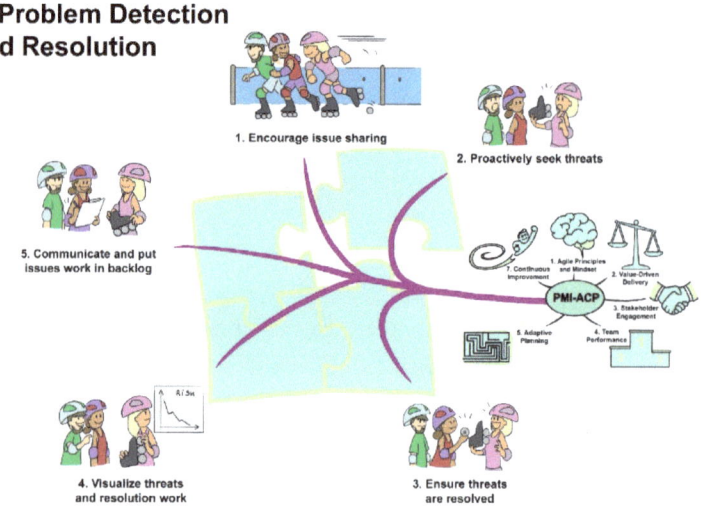

The Problem Detection and Resolution domain deals with threat and issue identification, threat resolution, and communications regarding issues and risks. The domain accounts for 10% of the exam questions, so expect around 12 questions on this topic in the exam.

Task 1 – Encourage issue sharing

Create a safe environment to talk about problems where no one gets criticized or blamed.

Catching things early greatly reduces the cost of changes, but that will not happen if people are reluctant to raise issues. We want people to feel confident and comfortable raising issues, so encourage honest dialogs and experimentation that can help improve performance.

Task 2 – Proactively seek threats

Teach people to be on the lookout for problems or issues at all times, then bring people's attention to them when they find any.

We need to proactively watch for threats and issues so we can prevent them from impacting us. When they are discovered we should try to find ways to improve the process to avoid or reduce any future impacts.

Task 3 – Ensure threats are resolved

Deal with problems quickly, or ensure people know about the issues that cannot be solved.

Fix the problem and communicate what happened or alert people to currently unfixed problems. It is bad enough for one person to step in a puddle; we don't want everyone else making the same mistake. So, fix it or warn others so they can avoid the problem.

Task 4 – Visualize threats and resolution work

Create a visible risk list to keep people focused on threats and aware of the work being done to overcome them.

Often things out of sight are out of mind. It is useful to keep a visual list of risks and issues to maintain awareness and focus on them. It also helps communicate priority, status, and ownership while encouraging action towards their resolution.

Task 5 – Communicate and put issues work in backlog

Put risk reduction work in the backlog of regular work.

Another strategy to ensure risk reduction work is not forgotten is to include it in the backlog of work. When threats are discovered, put the associated avoidance or reduction work in the backlog to provide better visibility into this work.

Domain 7 – Continuous Improvement

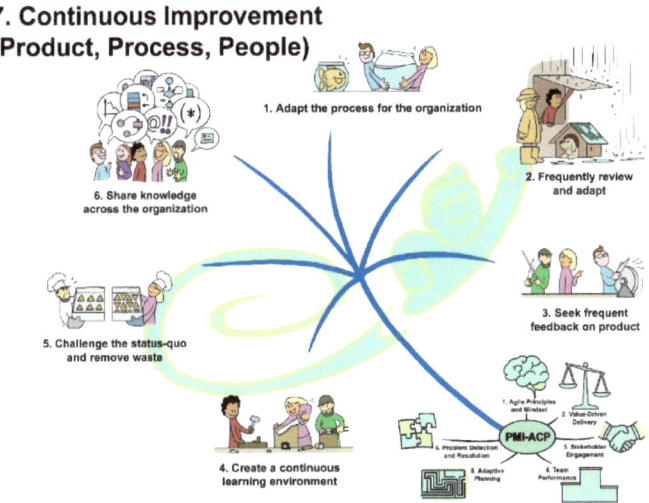

Domain 7 – Continuous improvement deals with the ongoing improvement of processes, tools, and people's capabilities. It accounts for 9% of the questions on the exam, so expect to see 10-12 questions on this topic in the test.

Task 1 – Adapt the process for the organization

Regularly adapting for growth and change at a large scale.

Periodically review how the team works and make changes to improve it. Things to evaluate include organizational culture, team practices, and delivery goals. We want to maximize the effectiveness of the team within acceptable norms and guidelines of the organization.

Task 2 – Frequently review and adapt

Frequently inspect and adapt through retrospectives.

Run experiments and make changes to improve our environment and effectiveness. Make sure we are not using retrospectives to passively record how things went but to actively shape future changes. We should be making improvements the effectiveness of individuals, teams, and the organization as a whole.

Task 3 – Seek frequent feedback on product

Ask for feedback on what you are building and incorporate that feedback to improve the product.

Regularly demonstrate the product being built and ask for feedback. Determine what the business and the customers value and what they do not value. Work to increase the value of the products and services being created.

Task 4 – Create a continuous-learning environment

Help people continue to develop their skills.

Provide learning opportunities for technical and interpersonal skills development. Encourage participation in growth and learning; not only do we reap the rewards on our project teams but grow and learn to contribute to people's ongoing happiness, sense of achievement, and success.

Task 5 – Challenge the status-quo and remove waste

Measure waste and value-adding work to improve the efficiency of how we work.

Do not assume things have to stay the same. Use techniques like value stream analysis to identify waste that can be removed to increase our effectiveness. Improve the processes being used to increase efficiency and competitiveness.

Task 6 – Share knowledge across the organization

Create processes for sharing learnings (good and bad) across the organization so that everyone benefits.

We want to make sure problems are not repeated but learnings and solutions are spread. Create structures and events to share information across project and department boundaries. Promote, fund, and invest in knowledge sharing.

Conclusion

Thanks for reading this book. I hope you found it helpful in explaining and illustrating these agile concepts. I enjoyed creating this lightweight, fun, and more graphical book. Please look out for more Agile Illustrated eBooks in the future.

Other Products

For anyone considering taking their PMI-ACP® exam, and why not, you have reviewed all the topics now – please check out my other PMI-ACP® products. Available here: https://bit.ly/2SqUKq5

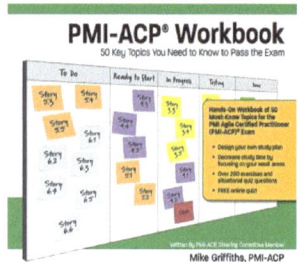

PMI-ACP Workbook – A summary of the core topics tested in the PMI-ACP® exam with exercises and practice questions.

Arranged alphabetically by topic, this book covers the top concepts in the PMI-ACP® exam with no additional fluff. Great for creating your own prioritized study plan, complete with Kanban board for tracking progress, this is for topic based learners.

PMI-ACP Exam Study Guide – A full text explanation of the agile mindset, values and principles tested in the exam. Along with the Kanban, Lean, Leadership and Emotional Intelligence topics also covered. All accompanied by review questions, exercises and sample exam questions.

I wrote this book to help people prepare for and pass their PMI-ACP® certification. Having worked on the design of the PMI-ACP® credential I knew the exam tested a multitude of concepts drawn from domains such as leadership, lean, kanban and agile approaches. So, I wanted to create a source that explains all the ideas in a single consistent voice.

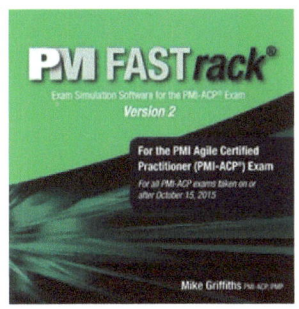

PMI-ACP FASTrack – Exam Simulator, know what to expect before you step into the exam room.

With 500+ questions the program allows you to filter questions by domain and keyword—and practice with randomized, timed 120 question tests with the same balance of questions from each domain as the real exam! All questions are also cross-referenced to the exam guide, so you may quickly and easily return to the book and work on your weak areas.

PMI-ACP Hot Topics - Spiral bound flashcards to help you study for the exam.

If you are looking for a way to prepare for the PMI-ACP® exam that fits into your busy schedule, these flashcards are it. Now you can study anywhere with RMC's portable and extremely valuable Hot Topics PMI-ACP® Exam Flashcards. This flip book offers over 175+ of the most important and difficult to recall PMI-ACP® exam-related terms and concepts—in portable format!

About the Author

Mike lives with his wife and son in Canmore, Alberta, Canada. He is an experienced agile coach with a long history of contributions to the agile community. Mike helped create the agile approach DSDM in 1994 and has been using agile approaches ever since. He served on the board of the Agile Alliance and remains active in the agile community.

Mike is a frequent contributor to the Project Management community. He was on the steering committee that created the Agile Certified Practitioner (PMI-ACP) credential and wrote RMC's ACP Exam Prep book. He was chairman of the author group for the PMI Agile Practice Guide, co-developed with the Agile Alliance. He also co-wrote the Software Extension to the PMBOK Guide and participates in the PMI Mentor program.

Mike trains and consults on effective project and product development. Combining elements of agile, leadership, and emotional intelligence, Mike tries to use the right tools for the job at hand. After 30 years of software project experience, Mike has come to realize that most complex projects can be simplified, and most conflicts can be resolved by taking the time to listen compassionately and jointly determine the best next steps. Projects do not need to be hard, but often in attempts to make things go quickly we make it harder than we need to.

Mike maintains the blog Leading Answers at www.LeadingAnswers.com. Check it for his most recent work and announcements.

www.ingramcontent.com/pod-product-compliance
Lightning Source LLC
Chambersburg PA
CBHW040318220526
45473CB00009B/2486